The Flock Book of Wensleydale Blue Faced Sheep
Volume 13

by Wensleydale Blue-faced Sheep Breeders Association

with an introduction by Jackson Chambers

Self Reliance Books

Get more historic titles on animal and stock breeding, gardening and old fashioned skills by visiting us at:

http://selfreliancebooks.blogspot.com/

Introduction

I am pleased to present yet another practical title on breeding and raising livestock.

The work is in the Public Domain and is re-printed here in accordance with Federal Laws.

As with all reprinted books of this age that are intended to perfectly reproduce the original edition, considerable pains and effort had to be undertaken to correct fading and sometimes outright damage to existing proofs of this title. At times, this task is quite monumental, requiring an almost total "rebuilding" of some pages from digital proofs of multiple copies. Despite this, imperfections still sometimes exist in the final proof and may detract from the visual appearance of the text.

I hope you enjoy reading this book as much as I enjoyed making it available to readers again.

Jackson Chambers

CONTENTS.

OFFICERS OF THE ASSOCIATION.

PRESIDENT.

Lord Henry Cavendish Bentinck, M.P.

VICE-PRESIDENTS.

Lord Ashton.

Mr. R. Burra, J.P.

Sir W. H. E. Chayter.

Mrs. Upton Cottrell-Dormer.

Mr. J. A. Farrer, J.P.

The Hon. W. H. Wentworth Fitzwilliam

Colonel W. H. Foster.

Mr. Francis Garth, J.P.

Mr. N. W. Helme, M.P.

Major E. G. S. Hornby, J.P.

The Hon. W. Lowther.

Mr. F. Punchard, J.P.

Mr. R. Rigg, M.P.

Mr. H. J. Storey.

Mr. F. W. Thompson, M.P.

Mr. T. Thompson, J.P.

MEMBERS OF COUNCIL.

Mr. F. Punchard, J.P. (Chairman).

Mr. R. Capstick.

Mr. J. Close.

Mr. J. Dargue.

Mr. A. Ewan.

Mr. R. Ewbank.

Mr. J. Handley, J.P.

Mr. A. Harker.

Mr. J. H. Holgate.

Mr. T. Jackson.

Mr. W. Millner.

Mr. J. Moore.

Mr. J. Percival.

Mr. W. Rhodes.

Mr. J. Scarr.

Mr. S. Thompson.

Mr. J. Towers.

Mr. J. Waller.

Mr. J. A. Willis.

Mr. T. Woof.

OBJECTS OF THE ASSOCIATION.

[EXTRACTED FROM THE MEMORANDUM OF ASSOCIATION].

(A) The encouragement of the breeding of Wensleydale Sheep at home and abroad, and the maintenance of their purity.

(B) The establishment and publication of a Flock Book of recognised and pure bred sires used in the past, and the annual registration of the pedigrees of such sires as are proved to the satisfaction of the Council to be eligible for entry.

(c) The annual compilation and publication of a statement of transactions connected with the breed, such as particulars relating to shows, sales and other usual transactions.

(D) The holding of shows and sales, the obtaining classes and giving prizes at various shows, and the appointment or recommendation of judges.

(E) The investigation of cases of doubtful and suspected pedigrees.

(F) The undertaking of the arbitration upon and settlement of disputes and questions relating to or connected with Wensleydale Sheep, and the breeding and sales thereof, and for other subsidiary purposes.

(G) Subject to the provisions of the 21st section of the Companies Act of 1862, the purchasing, hiring or taking on lease, for the purposes of the Society, any lands, houses or parts of houses, and the selling, letting and disposing of the property of the Society.

(H) The doing of all such other lawful things as are incidental or conducive to the attainment of the above objects or any of them.

REGULATIONS.

[EXTRACTED FROM THE ARTICLES OF ASSOCIATION].

II. MEMBERS.

4.—The subscribers of the Memorandum and Articles of Association, and the persons hereafter named as constituting the Council, shall be members of the Society, and any other persons desirous of becoming members of the Society, in accordance with these Articles, shall be proposed by one member of the Society and seconded by another member of the Society, and elected by a majority of the members present at a Council or General Meeting of the Society, and be entered on the register as such. Every person desiring to become a member shall, before being entered on the register, and becoming actually a member, and either before or after his election, sign a consent to become a member on a paper on which shall be printed or written clauses of these Articles or the regulations for the time being in force with respect to subscriptions.

5.—The rights and privileges of every member of the Society shall be personal to himself, and shall not be transferable or transmissible by his own act or by operation of law.

6.—Every member of the Society shall, provided he has paid his annual subscription for the current year, or has compounded for the same as hereinafter mentioned, be entitled to a copy of each of the Society's publications without further charge.

III. OBLIGATIONS OF MEMBERS.

7.—Every member on joining the Society shall pay to the funds of the Society an entrance fee, and shall also pay an annual subscription, or may, at his option, either on his joining the Society or at any other time afterwards, pay a lump sum by way of composition in lieu of such annual subscription; such entrance fee, subscription or composition to be according to a scale to be from time to time determined by the Council. The first annual subscription shall be considered as having fallen due on the 1st of January, 1894, and all subsequent annual subscriptions shall be payable in advance on the 1st day of January in each year, unless the member who would otherwise have been liable to pay the same shall give notice in writing to the Secretary before that date of his intention to withdraw from the Society.

8.—Every member shall observe all lawful bye-laws, regulations and orders of the Council, and pay all fines and forfeits which the Council shall in pursuance of their powers impose.

10.—Any member who shall neglect to pay any subscription, fine or forfeit for twelve calendar months consecutively may be excluded from the Society, and the removal of the name of such member from the register shall be sufficient evidence that he has been excluded by the Council, and he shall have no right of appeal or recision, provided always that this regulation shall not be construed to compel the Council to remove any member, or to give any member a right to be removed.

11.—The liability of a member who shall withdraw or be removed from the Society to pay any subscription, fine, or forfeit which has become due from him previous to his withdrawal or removal, shall not cease on his withdrawal or removal, but the Society may take such proceedings as shall be necessary for the recovery of such subscription, fine or forfeit.

31.—The President, Vice-Presidents, and Treasurer of the Society shall be *ex-officio* members of the Council if not elected members.

32.—All the members of the Council shall retire from office at the first annual meeting, and one-third of the members shall retire at each succeeding annual meeting, but be eligible for re-election. The members to retire each year after the first year shall, unless the Council agree among themselves, be determined by ballot.

33.—The quorum of the Council shall be five. The paraphrase " Special Council Meeting," in these Articles, shall mean a Council meeting of which not less than two clear days' notice is given to each member, specifying generally the nature of the business to be transacted.

34.—Two members of the Society shall be appointed by resolution at the annual meeting to be auditors of the then ensuing year.

35.—The Council, in special meeting, may supply any vacancy in the Council or officers which shall occur between one annual meeting and another, and the members or officers so appointed by the Council shall retire at the succeeding annual meeting.

36.—The books of account and accounts of the Society, with all the receipts and vouchers, shall, together with the annual statement of accounts, be delivered to the auditors for examination at least fourteen days before the day appointed for the annual meeting, and the auditors shall report thereon.

VII.   POWERS AND DUTIES OF COUNCIL.

37.—The management of the business of the Society shall be vested in the Council, who, in addition to the powers and authorities by statute or these Articles expressly conferred upon them, may exercise all such powers, and do all such acts and things as are or shall be, by statute or these Articles, directed or authorised to be done by the Society, and not hereby or by statute expressly directed to be done by the Society in general meeting assembled, but subject nevertheless to the statutory provisions and these Articles, and subject also to such (if any) regulations as may be from time to time determined by any special meeting of the Society, but no such regulation shall invalidate any prior act of the Council which would have been valid if the regulation had not been made.

BYE-LAWS.

The terms of membership shall be :—

(A) Life members, ten pounds.

(B) Annual subscribers who are not breeders, and members whose Flocks exceed ten ewes, an entrance fee of one guinea, and an annual subscription of half-a-guinea.

(c) *Bonâ fide* Tenant Farmers who have not more than ten ewes, an entrance fee of 10s. 6d., and an annual subscription of 10s. 6d.

(D) *Bonâ fide* Tenant Farmers who have not more than six ewes, and who joined the Society in or prior to the year 1897, an annual subscription of 2s. 6d.; and to those who join subsequently, an entrance fee of 5s. and an annual subscription of 5s.

The fee for entry of pedigrees shall be: members, 1s. for each ram; non-members, 5s. for each ram.

Export Certificates may be had on application to the Hon. Sec. Fee, 1s. each animal.

NOTE.—Every member can, on application to the Hon. Sec., obtain a printed copy of the Memorandum and Articles of Association on payment of a fee of 1s. for such copy. Every member will be deemed to have had full notice of all the clauses and conditions therein contained, and will be held bound by the same.

All letters and communications to be addressed to the Honorary Secretary,

Mr. W. RHODES,
Lundholme, Westhouse, near Kirkby Lonsdale.

STATEMENT OF ACCOUNTS FOR THE YEAR 1901.

RECEIPTS.

	£	s.	d.
Balance in Bank	65	12	11
Cash in hand	1	12	10
Annual Subscriptions from Vice-Presidents and Members	46	4	0
Entry Fees from New Members	6	5	0
Entry Fees for 82 Rams	4	2	0
Sundry Receipts for Advertisements, Inspectors' Expenses, Sale of Flock Books &c.	2	15	2
Interest from Debenture Stock, February and August	2	12	2
Bank Interest...	1	13	3
Entry Fees for Show and Sale ...	6	9	0
Entry Fees for Sale of Females ...	1	11	0
Cheques issued but not presented ...	1	3	3
	£140	0	7

EXPENDITURE.

	£	s.	d.
Mr. Hiscock's bill for Printing Vol. XII. of Flock Book, Advertising, Printing, &c., for Show and Sale... ...	19	17	1
Prize Money for Show and Sale ...	9	17	6
Donation to the Royal Lancashire Show	8	8	0
Mr. T. Otley for Medals	6	8	9
Hon. Sec. for Petty Cash, Stamps, &c. ...	3	10	0
Advertisement in *Farmer and Stockbreeder*	1	3	3
Mr. J. Brookes for Printing	2	17	1
Inspectors' Expenses...	2	1	3
Judges' Expenses and Refreshments, &c.	1	0	6
Cash in Bank	84	17	2
	£140	0	7

	£	s.	d.
Invested in Midland Debenture Stock ...	£100	0	0
Subscriptions not paid	4	8	6
Advertisements do.	1	4	0

Examined and found correct,

ROBERT F. KETTLEWELL,
J. C. CROFT,
} Auditors appointed by the Incorporated Wensleydale Blue-faced Sheep Breeders' Association.

RAMS

WHICH HAVE BEEN

USED IN REGISTERED FLOCKS.

---

685. **ALERT,**

Bred by Mr. E. Newhouse, Ancliffe Hall, Slyne, Lancaster;
lambed in 1898,
got by Falconer 297,
dam by Carbon 375,
gr. d. by Hematite 386.

Entered by E. Newhouse.

---

686. **APOLLO,**

Bred by Mr. J. Handley, Brigflatts, Sedbergh; lambed
in 1901,
got by Powder Blue 397,
dam by Scarbank 271,
gr. d. by Donald 23.

Entered by R. Redman, The Brows, Glasson Dock,
Lancaster.

687. **ASHBOURNE,**

Bred by Mr. T. Jackson, Netherbeck, Carnforth; lambed
in 1901,

got by Chamberlain 587,

dam by Cashier 327,

gr. d. by Rosario 172,

g. gr. d. by Warton 77.

Entered by Richard Parkinson, Bennett's Farm,
Preesall, Poulton-le-Fylde.

———

689. **ASHDALE PRINCE,**

Bred by Mr. F. E. C. Dobson, Dromonby House, Stokesley,
R.S.O., Yorkshire; lambed in 1897,

got by Sweetmeat 755,

dam by Faceby Bendigo,

gr. d. by Great Wonder.

Entered by the Exors. of the late T. Willis, The
Manor House, Carperby, Aysgarth, R.S.O.

———

690. **ASTERISK,**

Bred by Mr. W. Rhodes, Lundholme, Westhouse, Kirkby
Lonsdale; lambed in 1901,

got by Blue Beard 607,

dam by Wellington 236,

gr. d. by Swinethwaite 71,

g. gr. d. by Pluto 48,

g. g. gr. d. by Ajax 3.

Entered by G. Hitchon, Low Moor House, Clitheroe.

691. AUSTWICK,

Bred by Mr. Joseph Towers, Lawson's Farm, Nether
Kellet, Carnforth; lambed in 1898,
got by Lancaster Fashion 210,
dam by Dalesman 22,
gr. d. by Limestone Lad 389.

Entered by Mrs. Farrer, Ingleboro', Clapham,
Yorkshire.

---

692. BEST MAN,

Bred by Mr. J. Handley, Brigflatts, Sedbergh; lambed in
1901,
got by Beauty 603,
dam by Powder Blue 397,
gr. d. by Scarbank 271.

Entered by J. Towers, Lawson's Farm, Nether
Kellet, Carnforth.

---

693. BETTER LUCK,

Bred by the Exors. of the late T. Willis, The Manor
House, Carperby, Aysgarth, R.S.O.; lambed in 1900,
got by Royal Maidstone 582,
dam by Royal Manchester 458,
gr. d. by Heir of the Valley 259,
g. gr. d. by Aaron Arden 1,
g. g. gr. d. by St. Crispin 68,
g. g. g. gr. d. bred by the late Mr. T. Willis.

Entered by T. Sedgwick, Low Holme, Sedbergh.

694. ## BETTER STILL,

Bred by Mr. Redmayne Rigg, Wells House, Cartmel;
lambed in 1900,

got by Stainton 405,
dam bred by Mr. R. Rigg.

Entered by J. Gibson and Son, 168, Highgate,
Kendal.

---

695. ## BLUE BONNET,

Bred by Mr. J. Handley, Brigflatts, Sedbergh; lambed
in 1901,

got by Beauty 603,
dam by Powder Blue 397,
gr. d. by Scarbank 271.

Entered by J. Handley.

---

696. ## BLUE BOY,

Bred by Mr. R. Burra, Gate, Sedbergh, R.S.O.; lambed
in 1901,

got by Blue Skin 582,
dam by Marengo 499,
gr. d. by Swinethwaite 71,
g. gr. d. by Feudalist 26.

Entered by R. Burra.

697. # BLUE JACK,

Bred by the Exors. of the late T. Willis, Manor House,
Carperby, Aysgarth, R.S.O.; lambed in 1901,
got by Royal York 658,
dam by Ruler of the Valley 402,
gr. d. by Prospect Count 121,
g. gr. d. by Thorsby 72,
g. g. gr. d. by Westward Ho 79,
g. g. g. gr. d. bred by the late Mr. T. Willis.
Entered by J. Moore, Yorescott Farm, Askrigg,
R.S.O.

---

698. # BLUE KING,

Bred by Lord Henry Bentinck, M.P., Underley Hall, Kirkby
Lonsdale; lambed in 1901,
got by Bluebeard 607,
dam by Erl King 382,
gr. d. by True Blue 238,
g. gr. d. by Vulcan 76.
Entered by R. Capstick, Bramhaw, Sedbergh.

---

699. # BLUE PETER,

Bred by Mr. W. Rhodes, Lundholme, Westhouse, Kirkby
Lonsdale; lambed in 1901,
got by Marengo 499 or Welcome 593,
dam by Wellington 236,
gr. d. by Sir Peter 127,
g. gr. d. by Trojan 75,
g. g. gr. d. by Westward Ho 79,
g. g. g. gr. d. by Ajax 3.
Entered by R. Burra, Gate, Sedbergh, R.S.O.

700.   **BLUIT,**

Bred by Mr. J. Handley, Brigflatts, Sedbergh; lambed
in 1901,
    got by Beauty 608,
    dam by Powder Blue 397,
    gr. d. by Scarbank 271,

    Entered by J. Handley.

---

701.   **BOLTONIAN,**

Bred by Mr. E. Newhouse, Ancliffe Hall, Slyne, Lancaster;
lambed in 1900,
    got by Menestral 571.
    dam by Advancer 828,
    gr. d. by Viking 278.

    Entered by W. Dent and Sons, Street House,
Bolton, Penrith.

---

702.   **BONDSHOLME,**

Bred by Mr. J. Handley, Brigflatts, Sedbergh, R.S.O.;
lambed in 1901,
    got by Beauty 608,
    dam by Royal Victor 268,
    gr. d. by Sir William 64.

    Entered by W. Gibson, Broadrain Mill, Killington,
Sedbergh.

703. **BOWERSYKE,**

Bred by Mr. W. Gibson, Broadrain Mill, Killington,
Sedbergh ; lambed in 1901,
got by Successor 514,
dam by Stainton 405,
gr. d. by Professor 344,
g. gr. d. by Hopewell 101.

Entered by W. Gibson.

————

704. **BROUGHAM,**

Bred by Lord Henry Bentinck, M.P., Underley Hall,
Kirkby Lonsdale ; lambed in 1899,
got by Winder 523,
dam by Brough Sowerby 326,
gr. d. by Wellington 236.

Entered by T. Bainbridge, Brough Castle, Kirkby
Stephen.

————

705. **BRITISH RULER,**

Bred by Mr. E. Newhouse, Ancliffe Hall, Slyne, Lancaster ;
lambed in 1901,
got by Royal Ruler 459,
dam by Falconer 297,
gr. d. by Carbon 375,
g. gr. d. by Hematite 386.

Entered by J. Moffat, Rash Farm, Dent, Sedbergh,
R.S.O.

706.   CARNFORTH SWELL,

Bred by Mr. T. Jackson, Netherbeck, Carnforth ; lambed
in 1901,

got by Chamberlain 537,

dam by Cromwell 196,

gr. d. by Carnforth 13,

g. gr. d. by Favourite 25.

Entered by J. and T. Park, Bell Mount, Hest Bank,
Lancaster.

———

707.   CARPERBY,

Bred by Mr. James Close, Carperby, Aysgarth, R.S.O. ;
lambed in 1901,

got by King Edward 636,

dam by Royal Blue 350,

gr. d. by Wellington 236,

g. gr. d. by Sir Wilfred 70,

Entered by Joseph Thomas Camplin, Newby, near
Penrith.

———

708.   CASTERTON,

Bred by Mr. T. Jackson, Netherbeck, Carnforth ; lambed
in 1901,

got by Chamberlain 537,

dam by Conundrum 293,

gr. d. by Warton 77,

g. gr. d. by Favourite 25.

Entered by Thomas William Key, Old Hall Farm,
Casterton, Kirkby Lonsdale.

709.   **CAST OUT,**

Bred by Mr. J. Percival, East End House, Carperby,
Aysgarth, R.S.O.; lambed in 1901,

got by Ruler of the Valley 402,
dam by Dawning 379,
gr. d. by Hesperus 99,
g. gr. d. bred by the late Miss E. Willis.

Entered by Redmayne Rigg, Wells House, Cartmel.

———

710.   **CATHEDRAL,**

Bred by Mr. T. Jackson, Netherbeck, Carnforth ; lambed
in 1901,

got by Chamberlain 537,
dam by Cardinal 477,
gr. d. by Cromwell 196,
g. gr. d. by Carnforth 13.

Entered by T. Jackson.

———

711.   **CHARLES I.,**

Bred by Mr. A. Harker, Carperby, Aysgarth, R.S.O.;
lambed in 1887,

got by Woodland 80,
dam bred by Mr. J. Raw (and own sister to
Mr. J. Lambert's Royal York).

Entered by the Exors. of the late T. Willis, The
Manor House, Carperby, Aysgarth, R.S.O.

712.   **CHARLES II.,**

Bred by the late Mr. James Pickard, Thoresby, Aysgarth.
R.S.O.; lambed in 1890,
got by Lord of the Manor 108,
dam bred by the late Mr. Jas. Pickard.

Entered by the Exors. of the late T. Willis.

———

713.   **CLARE PRIDE,**

Bred by the Exors. of the late T. Willis, The Manor
House, Carperby, Aysgarth, R.S.O.; lambed in 1899,
got by Estimation 487,
dam by Sensation 353,
gr. d. by Sir James 58,
g. gr. d. by St. Crispin 68,
g. g. gr. d. bred by the late Mr. T. Willis.

Entered by R. Simpson, Manor Farm, Bletchley,
Bucks.

———

714.   **CLIFFORD PRINCE,**

Bred by Mr. R. Capstick, Bramhaw, Sedbergh ; lambed in
1901,
got by Brutus 474,
dam by Masterpiece 304,
gr. d. by Scargill 124,
g. gr. d. by Beatam 9.

Entered by S. Bargh, Clifford Hall, Burton-in-
Lonsdale, Kirkby Lonsdale.

715.   **CREDIT,**

Bred by Mr. T. Jackson, Netherbeck, Carnforth; lambed
in 1897,
got by Cashier 327,
dam by Rosario 172,
gr. d. by Warton 77.

Entered by J. Procter, Eccles Farm, Westhouse,
Kirkby Lonsdale.

———

716.   **DANDY JOE,**

Bred by Mrs. E. Cock and Sons, Red Bank, Bolton-le-
Sands, Carnforth; lambed in 1901,
got by Blood Royal 371,
dam by Clarence 425,
gr. d. by Herald 206.

Entered by G. Hitchon, Low Moor House, Clitheroe.

———

717.   **DANEGELT,**

Bred by Mr. Jos. Towers, Lawson's Farm, Nether Kellet,
Carnforth; lambed in 1894,
got by Dalesman 22,
dam by Bread Baker 372,
gr. d. by Limestone Lad 389.

Entered by F. Towers, Stubb Hall, Nether Kellet,
Carnforth.

718.   **DUFTON ROYAL,**

Bred by Mr. J. C. Croft, Carperby, Aysgarth, R.S.O.;
lambed in 1901,

got by Shepherd's Delight 659,

dam by Full of Fashion 429,

gr. d. by Ivy II.,

g. gr. d. by Castle Bank 248.

Entered by James Wills, Nether Hoff, Appleby,
Westmorland.

———

719.   **DUTIFUL BOY,**

Bred by Mr. J. Dargue, Beaumont Grange, Halton, Lancaster; lambed in 1898,

got by Owtsyde 450,

dam by Wellington 286,

gr. d. by Swinethwaite 71,

g. gr. d. by Pluto 48,

g. g. gr. d. by Ajax 3.

Entered by R. Winder, Townside Farm, Pilling,
Garstang.

———

720.   **EDEN BANK,**

Bred by Mr. W. Graham, Eden Grove, Kirkby Thore,
Penrith; lambed in 1901,

got by Westmorland 677,

dam by Brough Sowerby 326,

gr. d. by Son of Hesperus 99.

Entered by W. Dent and Sons, Street House,
Bolton, Penrith.

721.   EDENSIDE,

Bred by Mr. W. Graham, Eden Grove, Kirkby Thore,
Penrith ; lambed in 1898,

got by Duke 129,
dam by Victor 318,
gr. d. by Sir William 64,
g. gr. d. by Woodland 80.

Entered by W. Dent and Sons, Street House,
Bolton, Penrith.

---

722.   FOREMAN,

Bred by Mr. T. Jackson, Netherbeck, Carnforth ; lambed
in 1899,

got by Cardinal 477,
dam by Lord of the Valley 109,
gr. d. by Rosario 172,
g. gr. d. by Warton 77.

Entered by Thos. Fawcett, Holme, Carnforth.

---

723.   GAMESTER,

Bred by the Exors. of the late T. Willis, Manor House,
Carperby, Aysgarth, R.S.O. ; lambed in 1897,

got by Caractacus 374,
dam by William the Conqueror 184,
gr. d. by Thor's Hammer 73,
g. gr. d. by Thorsby 72,
g. g. gr. d. by St. Crispin 68,
g. g. g. gr. d. bred by the late T. Willis.

Entered by J. Waller, Low Thoresby, Aysgarth,
R.S.O.

724. GALLANT SUCCESSOR,

Bred by Mr. W. Gibson, Broadrain Mill, Killington,
Sedbergh; lambed in 1901,

got by Successor 514,
dam by Stainton 405,
gr. d. by Professor 344,
g. gr. d. by Donald II. 254.

Entered by J. Farrer, Abbot's Reading, Haverthwaite, Ulverston.

---

725. GIBSON,

Bred by Mr. W. Gibson, Broadrain Mill, Killington,
Sedbergh; lambed in 1901,

got by Successor 514,
dam by Stainton 405,
gr. d. by Donald II. 254.

Entered by J. Handley, Brigflatts, Sedbergh.

---

726. HIS MAJESTY,

Bred by Mr. A. Ewan, Gooda, Westhouse, Kirkby Lonsdale; lambed in 1901,

got by Grand Quality 630,
dam by Yorkshire Fashion 413,
gr. d. by Herald 206,
g. gr. d. by Thorsby II. 135.

Entered by A. Ewan.

727.   **HELMSIDE,**

Bred by Mr. R. Johnson, Cragges Farm, Dent, Sedbergh ;
lambed in 1901,

     got by Signet 660,
     dam by Conundrum 298,
     gr. d. by Beatam 9,
     g. gr. d. by Golden Locks 96.

Entered by R. Johnson.

---

728.   **JESTER,**

Bred by Lord Henry Bentinck, M.P., Underley Hall,
Kirkby Lonsdale ; lambed in 1901,

     got by Welcome 598,
     dam by William Victor 467.

Entered by Anthony Robson, Carperby, Aysgarth,
R.S.O.

---

729.   **KELLET II.,**

Bred by Mr. J. Handley, Brigflatts, Sedbergh ; lambed in
1901,

     got by Powder Blue 897,
     dam by Kellet 490,
     gr. d. by Donald 23.

Entered by J. Handley.

730. **KEEPSAKE,**

Bred by Mr. W. Graham, Eden Grove, Bolton, Penrith;
lambed in 1899,

got by Nansen 391,

dam by Compensation,

gr. d. by Eton 548.

Entered by W. Graham.

———

731. **LEADING STAR,**

Bred by Messrs. W. and R. Millner, Slyne Hall, near
Lancaster; lambed in 1901,

got by Harrogate Lad 433,

dam by Royal Ruler 459,

gr. d. by Advancer 323,

g. gr. d. by Regulator 222,

g. g. gr. d. by a ram bred by the late Miss
E. Willis.

Entered by R. Millner, Bolton-le-Sands, Carnforth.

———

732. **LITTLE DICKY,**

Bred by Mr. J. Ireland, Bulk, Lancaster; lambed in 1892,

got by Reserve,

dam by Satisfaction.

Entered by T. Walker, Templand, Grange-over-
Sands.

733.   **LONG TOM,**

Bred by Mr. J. Rhodes, Stockeld, Wetherby, lambed in
1900,

got by Gamester 723,
dam by Stockeld II.

Entered by J. Waller, Low Thoresby, Aysgarth,
R.S.O.

---

734.   **LORD KITCHENER,**

Bred by Mr. T. Kitchen, Tatham Hall, Wray, Lancaster ;
lambed in 1901,

got by Bellmontina 604,
dam by Modderina 574,
gr. d. by Sir James 403,
g. gr. d. by Baronet 8.

Entered by N. Newton, senr., Overtown Farm,
Kirkby Lonsdale.

---

735.   **LYVENNET,**

Bred by Messrs. W. Dent and Sons, Street House, Bolton,
Penrith ; lambed in 1901,

got by Boltonian 701,
dam by Mikado 788,
gr. d. by Prime Minister 744.

Entered by W. Dent and Sons.

736.   **MARQUIS,**

Bred by Mr. T. Thompson, Prospect House, Hest Bank,
Lancaster ; lambed in 1901,
got by Duke of Lancaster 623,
dam by Viking 278,
gr. d. by Falconer 297,
g. gr. d. by Dandy Pat 197.

Entered by Anthony Harker, Carperby, Aysgarth,
R.S.O.

---

737.   **MEDALIST,**

Bred by Mr. J. Percival, East End House, Carperby,
Aysgarth, R.S.O. ; lambed in 1901,
got by Ruler of the Valley 402,
dam by Honesty 207,
gr. d. by Lord of the Manor 108,
g. gr. d. by Feudalist 26.

Entered by J. Percival.

---

738.   **MIKADO,**

Bred by Mr. L. T. Pickering, Bourn, Cambridge ; lambed
in 1893,
got by True Blue II. 277,
dam by a Son of Harbinger 32,
gr. d. by a Son of Sir Wilfred 63,
g. gr. d. by a ram bred by Mr. J. Pickard.

Entered by W. Dent and Sons, Street House,
Bolton, Penrith.

739. **NORTHERN LIGHT,**

Bred by Mr. W. Rhodes, Lundholme, Westhouse, Kirkby
Lonsdale ; lambed in 1901,

    got by President 578,

    dam by Viking 278,

    gr. d. by Trojan 75,

    g. gr. d. by Pluto 48,

    g. g. gr. d. by Ajax 8.

Entered by J. Hargreaves, Lane Side, Sawley, near
Clitheroe.

---

740. **NEW BOY,**

Bred by Mr. J. Whitwell, Holden House, Westhouse, near
Kirkby Lonsdale ; lambed in 1901,

    got by Successor 514,

    dam by Stainton 405,

    gr. d. by William the Conqueror 184,

    g. gr. d. by Sir James 58.

Entered by J. Whitwell.

---

741. **PETER TEAZLE,**

Bred by Mr. Richard Procter, Barkerfield, Downham,
Clitheroe ; lambed in 1901,

    got by Maximillian 645,

    dam by Proud Pendle 452,

    gr. d. by Baronet 8,

    g. gr. d. by Westward Ho 79.

Entered by R. Procter.

742. PRESIDENT BROUGH,

Bred by Mr. W. Rhodes, Lundholme, Westhouse, Kirkby
Lonsdale ; lambed in 1901,

got by President 578,
dam by Safeguard 270,
gr. d. by Sir Peter 127,
g. gr. d. by Excelsior 24,
g. g. gr. d. by Baron Bolton 7.

Entered by T. Bainbridge, Brough Castle, Kirkby
Stephen.

---

743. PRIDE OF NETHERBECK,

Bred by Mr. T. Jackson, Netherbeck, Carnforth ; lambed
in 1901,

got by Cardinal 477,
dam by Cashier 327,
gr. d. by Rosario 172,
g. gr. d. by Warton 77.

Entered by Henry Dawson Greene, Whittington
Hall, Kirkby Lonsdale.

---

744. PRIME MINISTER,

Bred by Mr. Robert Burra, Gate, Sedbergh, R.S.O.; lambed
in 1892,

got by Baronet 8,
dam by Excelsior 24,
gr. d. by Pluto 48,
g. gr. d. bred by Mr. R. Capstick, Bramhaw,
Sedbergh.

Entered by W. Dent and Sons, Street House, Bolton,
Penrith.

745.   PRINCE ROBERTUS,

Bred by Mr. J. W. Fothergill, Brownber, Newbiggin, R.S.O.,
Westmorland ; lambed in 1901,

got by General Roberts 556,

dam by Masterful 303,

gr. d. by True Blue 233,

g. gr. d. by Feudalist 26.

Entered by Richard Hutchinson, Brough Sowerby
Lodge, Kirkby Stephen.

———

746.   PROSPECT KING,

Bred by Mr. T. Thompson. Hest Bank, Lancaster ; lambed
in 1901,

got by Duke of Lancaster 623,

dam by Lord of the Valley 109,

gr. d. by Thorsby 72,

g. gr. d. by St. Crispin 68.

Entered by T. Thompson.

———

747.   PROSPECT VIKING,

Bred by the Exors. of the late T. Willis, The Manor House,
Carperby, Aysgarth, R.S.O. ; lambed in 1900,

got by Estimation 487,

dam by Caractacus 374,

gr. d. by William the Conqueror 184,

g. gr. d. by Thor's Hammer 78,

g. g. gr. d. by Westward Ho 79,

g. g. g. gr. d. bred by the late Mr. T. Willis.

Entered by T. Thompson, Prospect House, Hest
Bank, near Lancaster.

748.   **ROYAL BOY,**

Bred by Mr. G. Whittaker, Gibson's Farm, Kirkland, Garstang ; lambed in 1901,
got by Yorkshire Fashion 413,
dam by Dark Blue 483.
Entered by G. Whittaker.

---

749.   **ROB ROY,**

Bred by Messrs. W. and R. Millner, Slyne Hall, Lancaster ; lambed in 1901,
got by Harrogate Lad 433,
dam by Master Recorder 446,
gr. d. by Advancer 328,
g. gr. d. by Regulator 222,
g. g. gr. d. by Dalesman 22,
g. g. g. gr. d. by a ram bred by Mr. R. Capstick.
Entered by J. Hoggarth, Manor House, Slyne, Lancaster.

---

750.   **ROYAL CARDIFF,**

Bred by the Exors. of the late Mr. T. Willis, The Manor House, Carperby, Aysgarth, R.S.O. ; lambed in 1900,
got by Royal Maidstone 582,
dam by Sensation 353,
gr. d. by Lord of the Valley 109,
g. gr. d. by Thorsby 72,
g. g. gr. d. by St. Crispin 68,
g. g. g. gr. d. bred by the late Mr. T. Willis.
Entered by the Exors. of the late T. Willis.

751.   **ROYAL COUNT,**

Bred by Mr. T. Thompson, Prospect House, Hest Bank,
Lancaster ; lambed in 1898,
got by Royal Premier 352,
dam by Lord of the Valley 109,
gr. d. by Thorsby 72,
g. gr. d. by St. Crispin 68.
Entered by T. Thompson.

---

752.   **ROYAL DUKE,**

Bred by Messrs. W. and R. Millner, Slyne Hall, Lancaster ; lambed in 1901,
got by Blood Royal 371,
dam by Dandy Pat 197.
Entered by W. and R. Millner.

---

753.   **ROYAL SCOTSMAN,**

Bred by Mr. J. Chapman. Leyburn ; lambed in 1900,
got by Ashdale Prince 689,
dam by Charles II. 712,
gr. d. by Charles I. 711.
Entered by the Exors. of the late T. Willis, The
Manor House, Carperby, Aysgarth, R.S.O.

---

754.   **SIR WILFRED,**

Bred by Mr. G. Whittaker, Gibson's Farm, Kirkland,
Garstang ; lambed in 1901,
got by Yorkshire Fashion 413,
dam by Dark Blue 483.
Entered by G. Whittaker.

755. SWEETMEAT,

Bred by Mr. E. Alderson, Leyburn ; lambed in 1891,
got by Bellerby 85,
dam by a ram of Mr. G. Armistead's.
Entered by the Exors. of the late T. Willis, The
Manor House, Carperby, Aysgarth, R.S.O.

---

756. SUCCESSOR II.

Bred by Mr. J. Handley, Brigflatts, Sedbergh ; lambed in
1901,
got by Successor 514,
dam by Scarbank 271,
gr. d. by Donald 23.
Entered by J. Handley.

---

757. SURPRISE,

Bred by Mr. J. Handley, Brigflatts, Sedbergh ; lambed in
1901,
got by Beauty 603,
dam by Powder Blue 397,
gr. d. by Scarbank 27.
Entered by J. Handley.

---

758. TEMPLAND BOY,

Bred by Mr. T. Jackson, Netherbeck, Carnforth ; lambed
about 1888,
got by Carnforth 13,
dam by Favourite 15.

Entered by T. Walker, Templand, Grange-over-
Sands.

759. **TEMPLAND CHIEF,**

Bred by Messrs. J. Gibson and Son, Kendal; lambed in
1900,

got by Stainton 405,

dam by Natland Chief 390,

gr. d. by Real Blue 267,

g. gr. d. by Hopeful 33.

Entered by T. Walker, Templand, Grange-over-
Sands.

---

760. **TEMPLAND GEM,**

Bred by Mr. J. Dargue, Bee Nest, Casterton, Kirkby
Lonsdale; lambed in 1896,

got by Viking 278,

dam by Wellington 236,

gr. d. by Trojan 75,

g. gr. d. bred by the late C. Duckett.

Entered by T. Walker, Templand, Grange-over-
Sands.

---

761. **TEMPLAND SWELL,**

Bred by Mr. S. Bargh, Clifford Hall, Burton-in-Lonsdale,
Kirkby Lonsdale; lambed in 1900,

got by Welcome 593,

dam by Chancellor 89,

gr. d. bred by Mr. T. Newton.

Entered by T. Walker, Templand, Grange-over-
Sands.

762.   **TOP SAWYER,**

Bred by Mr. W. Rhodes, Lundholme, Westhouse, Kirkby
Lonsdale ; lambed in 1901,

got by Marengo 499,

dam by Erl King 382,

gr. d. by Baronet 8,

g. gr. d. by Ajax 3.

Entered by the Exors. of the late J. Walsh, Bourne
Hall, Poulton-le-Fylde.

———

763.   **THE HERO,**

Bred by Mr. W. Rhodes, Lundholme, Westhouse, Kirkby
Lonsdale ; lambed in 1901,

got by Welcome 593,

dam by Mint Rock 305,

gr. d. by Beatam 9,

g. gr. d. by Lord John 40.

Entered by James Close, Chapel House, Carperby,
Aysgarth, R.S.O.

———

764.   **THE SIRDAR,**

Bred by Mr. W. Gibson, Broadrain Mill, Sedbergh ; lambed
in 1897,

got by Stainton 405,

dam by Donald II. 254,

gr. d. by Newbiggin 217.

Entered by W. Dent and Sons, Street House, Bolton,
Penrith.

765. **THE SWELL,**

Bred by Mr. A. Harker, Carperby, Aysgarth, R.S.O.;
lambed in 1901,

got by Councillor 620,
dam by Beaumont Chief 325,
gr. d. by Bellerby 85,
g. gr. d. by Hesperus 99.

Entered by Jas. Wills, Nether Hoff, Appleby, Westmorland.

---

766. **WAGTAIL,**

Bred by Mr. J. H. Holgate, Foxley Bank, Grindleton,
Clitheroe; lambed in 1901,

got by Wellspring 675,
dam by Erl King 582.

Entered by J. H. Holgate.

---

767. **WHITTINGTON,**

Bred by Mr. H. Dawson Greene, Whittington Hall, Kirkby
Lonsdale; lambed in 1901,

got by Prince Harold 579,
dam by Boy in Blue 472,
gr. d. by Viking 278,
g. gr. d. by Excelsior 24,
g. g. gr. d. by Baron Bolton 7.

Entered by R. Capstick, Bramhaw, Sedbergh.

768.   WHITTINGTON HERO,

Bred by Mr. H. Dawson Greene, Whittington Hall, Kirkby
Lonsdale ; lambed in 1901,

got by Prince Harold 579,

dam by Stainton 405,

gr. d. by Kent Hero 105,

g. gr. d. by Hopeful 33.

Entered by H. Dawson Greene.

———

769.   WINTER,

Bred by Mr. W. Rhodes, Lundholme, Westhouse, Kirkby
Lonsdale ; lambed in 1901,

got by Welcome 593,

dam by Marengo 499,

gr. dam by Wellington 236,

g. gr. d. by Swinethwaite 71,

g. g. gr. d. by Pluto 48,

g. g. g. gr. d. by Ajax 3.

Entered by J. H. Holgate, Foxley Bank, Grindleton,
Clitheroe.

———

770.   YEOMAN,

Bred by Mr. J. Handley, Brigflatts, Sedbergh ; lambed in
1901,

got by Successor 514,

dam by Powder Blue 397,

gr. d. by Scarbank 271.

Entered by T. Thompson, Prospect House, Hest
Bank, Lancaster.

PARTICULARS OF THE BREEDING

OF

REGISTERED FLOCKS in 1901.

ASHTON, LORD,

Ryelands, Lancaster.

See Vol. XII., page 29.

Rams used in 1901: Burrow Premier 475, Burrow Reliance 615, and Burrow Prince 614.

ABBOT BROS.,

Thuxton, Norfolk.

See Vol. XII., page 29.

Rams used in 1901: Yarmouth 595 and Yarrow 596.

BAINBRIDGE, THOS.,

Brough Castle, Kirkby Stephen.

Ten ewes put to the ram.

This flock has been in existence for 20 years, being founded from sheep purchased from Messrs. Kettlewell, R. Bell, and Fawcett, Askrigg; and the rams used have been selected with great care and bred by the late Mr. T. Willis, Mr. G. Bell, Askrigg, Mr. J. W. Fothergill, and in 1899 Brougham 704 (bred by Lord Henry Bentinck, M.P.) was purchased and used for two seasons in the flock.

Ram used in 1902: President Brough 742.

———

BANKS, J. HENRY,

Trunnah Farm, Thornton, Poulton-le-Fylde.

See Vol. XII., page 29.

Eight ewes put to the ram.

Ram used in 1901: Top Sawyer 762.

Prizes won in 1901 :—

LYTHAM.—3rd, Three Gimmer Lambs.

POULTON-LE-FYLDE AUCTION MART.—1st and 2nd, Ram Lamb.

———

BANKS, THOMAS,

Bilsborough, Preston, Lancashire.

See Vol. XII., page 30.

Thirteen ewes put to the ram.

Ram used in 1901 : Erl King 382.

Prize won in 1901 :—

GOOSNARGH SHOW.—2nd, Three Gimmer Lambs.

BARGH, SAMUEL,

Clifford Hall, Burton-in-Lonsdale, Kirkby Lonsdale.

See Vol. XII., page 30.

Seven ewes put to the ram.

Ram used in 1901 : Clifford Prince 714.

———

BARKER, JOSEPH,

Maythorne, Leyburn.

See Vol. XII., page 30.

Twenty-nine ewes put to the ram.

Rams used in 1901 : Macdonald 642, Royal Scotsman
758.

———

BATEMAN, JOHN,

Hall Garth, Clapham, Yorkshire.

See Vol. XII., page 31.

Six ewes put to the ram.

Ram used in 1901 : Edgar 547.

———

BLEASDALE, JOSEPH,

Pump House, Nether Kellet, Carnforth.

See Vol. XII., page 31.

Twenty ewes put to the ram.

Rams used in 1901 : Beaumont Beauty 530, Mafeking
569 and Alert 685.

BENTINCK, LORD HENRY, M.P.,

Underley Hall, Kirkby Lonsdale.

See Vol. XII., page 31.

Forty-five ewes put to the ram.

In 1901 three ewes were added bred by Mr. G. Hitchon, Clitheroe; two by Waller 363 and one by Mint Rock 305, also two from Mr. R. Burra, Gate, Sedbergh, by Recollection 453 and Woodman 525.

Rams used in 1901: Welcome 593, Blue Beard 607 and Carnegie 618.

Prizes won in 1901 :—

ROYAL SHOW AT CARDIFF.—Reserve and h.c., Shearling Ram; Reserve and h.c., Three Shearling Ewes.

ROYAL LANCASHIRE SHOW AT ST. HELENS.—2nd and 3rd, Ram, two-shear and over; 3rd, Ram Shearling; 3rd, Three Shearling Ewes; 1st, Ram Lamb; 3rd, Three Ewe Lambs.

YORKSHIRE AGRICULTURAL SHOW AT BRADFORD.—2nd, Ram any age.

PENRITH SHOW.—1st, Ram, more than one-shear; 2nd, Shearling Ram; 1st, Pair of Shearling Gimmers; 1st and 2nd, Ram Lamb; 1st, Pair of Gimmer Lambs.

LANCASTER.—1st, Ram, two-shear and upwards; 1st Shearling Ram; 1st and 2nd, Three One-shear Gimmers; 2nd, Three Ewes.

BURTON, MILNTHORPE AND CARNFORTH SHOW AT MILNTHORPE.—1st, Ram, any age; 1st Three Ewes; 1st and 3rd, Three One-shear Gimmers; 3rd, Ram Lamb.

SETTLE SHOW.—1st and 2nd, Ram Lamb; 1st, Shearling Ram to cross with white-faced or half-bred ewes; 1st, Ram of any age; 1st and 2nd, Three One-shear Gimmers; 1st, Three Stock Ewes; 2nd, Shearling Ram to cross with a Scotch Ewe; 1st, Special Prize given by J. A. Farrer, Esq., for best Teeswater, Wensleydale, or Leicester.

LUNESDALE SHOW.—1st and 2nd, Ram Lamb; 1st, Three Gimmer Lambs; 1st and 2nd, Shearling Ram; 1st and 2nd, Aged Ram; 1st and 2nd, Three One-shear Gimmers; 1st, 2nd and 3rd, Three Stock Ewes; Silver Medal, Special Prize given by the Wensleydale Sheep Breeders' Association for Pair of Lambs (either sex).

WESTMORLAND AND KENDAL SHOW.—1st, Aged Ram; 3rd, Shearling Ram; 2nd, Three Ewes; 1st and 2nd, Three Shearling Gimmers; 2nd, Three Gimmer Lambs; 1st, Special Prize given by Lord Henry Bentinck, M.P., for Collection of Wensleydale Sheep, Three Ewes, Three Shearlings, Three Lambs and Aged Ram.

KENDAL AUCTION MART.—1st, Wensleydale Ram.

BURRA, R.,

Gate, Sedbergh, R.S.O.

See Vol. XII., page 33.

Fourteen ewes put to the ram.

Rams used in 1901: Erl King II. 485, Blue Peter 699, Blue Boy 696 and British Ruler 705.

Prizes won in 1901:—

SEDBERGH.—2nd, Aged Ram; 2nd, Three Shearling Gimmers; 1st, Three Ewes.

CAMPLIN, J. T.,

Newby, Penrith.

See Vol. XII., page 33.

Nine ewes put to the ram.

Ram used in 1901 : Carperby 707.

———

CAPSTICK, RICHARD,

Bramhaw, Sedbergh.

See Vol. XII., page 34.

Fifty-five ewes put to the ram.

Rams used in 1901 : Blue King 698 and Whittington 767.

Prizes won in 1901 :

SEDBERGH.—2nd, Stock Ewes ; 2nd, Gimmer Lambs ; 2nd, Shearling Ewes ; 3rd, Tup Lambs.

SOCIETY'S SHOW AND SALE AT HELLIFIELD.—1st and c., Ram Lamb.

———

CAPSTICK and SON,

Moser Hill, Dent, Sedbergh.

See Vol. XII., page 34.

Six ewes put to the ram.

Ram used in 1901: Wandering Boy 671.

CLOSE, JAMES,

Carperby, Aysgarth Station, R.S.O.

See Vol. XII., page 34.

Ten ewes put to the ram.

Ram used in 1901 : The Hero 768.

Prizes won in 1901 :—

LEYBURN.—2nd, Shearling Ram ; 1st, Three Shearling Ewes.

HAWES.—1st, Three Ewes ; 1st, Three Gimmer Lambs ; 2nd, Two Ram Lambs.

COCK, E., and SONS,

Red Bank, Bolton-le-Sands, Carnforth.

See Vol. XII., page 35.

Twenty-eight ewes put to the ram.

One ewe added from E. S. Jackson, M.B., Carnforth ; sire Owtsyde 450, dam by Falconer 297, gr. d. by Dandy Pat 197.

Rams used in 1901 : Blood Royal 871 ; Advancer 828 ; Rob Roy 749 and Prospect King 746.

CROFT, J. CARTER,

Carperby, Aysgarth Station, R.S.O.

See Vol. XII., page 35.

Six ewes put to the ram.

Rams used in 1901 : Shepherd's Delight 659 and Marquis 736.

DARGUE, JOHN,

Bee Nest, Casterton, Kirkby Lonsdale.

See Vol. XII., page 35.

Eighteen ewes put to the ram.

One ewe added bred by Mr. James Gibson, Kendal; got by Successor 514, dam by Stainton 405, gr. d. by Natland Swell 307, by Exchequer 202.

Ram used in 1901: Welcome 598.

DENT, W. and SONS,

Street House, Bolton, Penrith.

Twenty-six ewes put to the ram.

In 1892 six shearlings were purchased from Mr. Thomas Bainbridge, Brough Castle, Westmorland; got by a ram bred by the late T. Willis, Manor House, Carperby. Their dams were from the flock of the late Mr. Kettlewell. In 1893 four shearlings were added from the flock of Mr. Jas. Close, Carperby; got by Dalesman 211, and their dams were from Mr. Pilkington's flock. In 1895 five shearlings were added from the flock of Mr. R. Burra, Gate, Sedbergh; got by Wellington 236. In 1898 two were added from the flock of Mr. J. Fothergill, Brownber; got by Royal Dalesman 509. In 1900 two were added from the flock of Mr. T. Thompson, Prospect House, Hest Bank; got by Vulcan 76; also two from the flock of Mr. J. Percival, Carperby, by Daybreak 484.

Ram used in 1892 : Prime Minister 744.

   ,,   1893 : Prime Minister 744, Mikado 738.

   ,,   1894 : Prime Minister 744, Mikado 738.

   ,,   1895 :   do.   do.

   ,,   1896 : Mikado 738.

   ,,   1897 : Mikado 738 and The Sirdar 764.

   ,,   1898 : The Sirdar 764, Edenside 721.

   ,,   1899 : The Sirdar 764, Woodman 525.

   ,,   1900 : The Sirdar 764, Woodman 525, Craven Bank 542, Boltonian 701.

   ,,   1901 : Craven Bank 542, Boltonian 701. Eden Bank 720, Lyvennet 735.

DICKENSON, JOHN,

Pit Farm, Grange-over-Sands.

Six ewes put to the ram.

This flock was commenced in 1900 by purchasing from Mr. J. Gibson, Natland Hall, Kendal, six ewes, two of which were by Natland Chief 390, two by Stainton 405, and one each by Wharton 77 and Natland Swell 307.

   Rams used in 1900 : Brigflatts Blue 682 and Dainty Stamp 621.

     ,,   1901 : Brigflatts Blue 682.

DINSDALE, R.,

Redmire, Leyburn, R.S.O.

See Vol. XII., page 86.

Six ewes put to the ram.

Ram used in 1901 : Long Tom 733.

DINSDALE, WILLIAM,

Carperby, Aysgarth Station, R.S.O.

See Vol. XII., page 86.

One ewe put to the ram.

Ram used in 1901: Royal Cardiff 750.

———

EWAN, AARON,

Gooda, Westhouse, Kirkby Lonsdale.

See Vol. XII., page 86.

Thirty ewes put to the ram.

Rams used in 1901: Grand Quality 680 and His
Majesty 726.

Prizes won in 1901:—

BENTHAM SHOW.—1st and Special for Shearling Ram,
GRAND QUALITY 680.

SHOW AND SALE AT HELLIFIELD.—2nd for Shearling Ram,
THORSBY IV. 667.

———

EWBANK, ROBERT,

Lawkland Green, Clapham, Yorkshire.

See Vol. XII., page 86.

Six ewes put to the ram.

Rams used in 1901: Lawkland I. 638 and Lawkland
II. 689.

Prizes won in 1900 :—

SETTLE.—1st, Ram Lamb ; 2nd, Shearling Ram ; 1st
and 2nd, Shearling Rams for crossing purposes ; 1st,
Special for Three Best Tups in Show.

SKIPTON.—1st, Ram Lamb ; 2nd, Shearling Ram.

SOCIETY'S SHOW AND SALE AT HELLIFIELD.—2nd, Aged
Ram.

FARRER, MRS.,

Ingleboro' Hall, Clapham.

See Vol. XII., page 37.

Eight ewes put to the ram.

Rams used in 1901: Lawkland I. 638, Lawkland II. 689 and Austwick 691.

FARRER, JOHN,

Abbot's Reading, Haverthwaite, Ulverston.

See Vol. XII., page 37.

Ten ewes put to the ram.

Ram used in 1901: Reading Blue 655 and Gallant Successor 724.

Prizes won in 1901:—

NORTH LONSDALE SHOW AT ULVERSTON.—1st and 3rd, Ram Lamb; 2nd, Three Gimmer Lambs.

HAWKSHEAD.—2nd and 3rd, Shearling Ram; 2nd and 3rd, Ram Lamb; 1st and 2nd, Gimmer Lambs; 1st, Gimmer Shearlings; 1st, Ewes; Special, Best Female Sheep.

LOWICK.—1st, Gimmer Lambs; 3rd, Gimmer Shearling.

FAWCETT, THOS.,

Holme, Carnforth.

Six ewes put to the ram.

Ram used in 1901: Foreman 722.

FITZWILLIAM, HON. W. H. W., M.P.,

Wigganthorpe, York.

See Vol. XII., page 88.

Forty-four ewes put to the ram.

Ram used in 1901: Ribblesdale Enterprise 580.

FOTHERGILL, J. W.,

Brownber, Newbiggin, R.S.O., Westmorland.

See Vol. XI., page 88.

Fifteen ewes put to the ram.

Four ewes were added in 1901 from Mr. Ewan, also three ewes from the flock of Dr. R. W. Gibson.

Ram used in 1901: General Roberts 556.

GIBSON, JAMES,

163, Highgate, Kendal.

See Vol. XII., page 88.

Twenty ewes put to the ram.

In 1901 five ewes were added from Mr. W. Gibson, Broadrain Mill, and four ewes from Mr. Redmayne Rigg, Cartmel; also three gimmer lambs from Mr. R. Capstick, Sedbergh.

Rams used in 1901: Dainty Stamp 621 and Better Still 694.

GIBSON, R. W.,

Orton, Tebay, Westmorland.

See Vol. XII., page 38.

Ten ewes put to the ram.

Ram used in 1901: Viscount 519.

GIBSON, WILLIAM,

Broadrain Mill, Killington, Sedbergh.

See Vol. XII., page 39.

Nineteen ewes put to the ram.

Rams used in 1901: Bowersyke 708 and Bondsholme 702.

GRAHAM, WILLIAM,

Eden Grove, Bolton, Penrith.

See Vol. XII., page 39.

Twenty-five ewes put to the ram.

Rams used in 1901: Kingscraft 635, Keepsake 730.

GREENE, H. DAWSON,

Whittington Hall, Kirkby Lonsdale.

Fifteen ewes put to the ram.

Rams used in 1901: Prince Harold 579, Pride of Netherbeck 748 and Whittington Hero 768.

Prizes won in 1901:

SETTLE.—1st, Three Gimmer Lambs.

BENTHAM.—1st, Aged Ram; 1st, Three Gimmer Lambs; 1st, Three Ewes; 2nd, Three Shearling Gimmers; 2nd Ram Lamb; 3rd, Ram Lamb.

KIRKBY LONSDALE.—2nd, Three Gimmer Lambs.

Above were only times shown.

HANDLEY, JOHN,

Brigflatts, Sedbergh.

See Vol. XII., page 40.

Thirty-three ewes put to the ram.

Rams used in 1901: Powder Blue 897, Beauty 608, Gibson 725, Estimation 487, Prospect Viking 747 and Bondsholme 702.

Prizes won in 1901:—

ROYAL LANCASHIRE AT ST. HELENS.—2nd for Gimmer Lambs; 3rd for Tup Lamb.

LANCASTER SHOW.—2nd, Three Tup Lambs (sale class); 1st, President's Special for One Male and Three Females.

LUNESDALE SHOW AT KIRKBY LONSDALE.—2nd, Best Pair of Wensleydale Lambs given by the Wensleydale Sheep Breeders' Association.

BURTON, MILNTHORPE AND CARNFORTH SHOW.—2nd, Three Gimmer Shearlings; 2nd, Three Gimmer Lambs; 2nd, Three Tup Lambs; 1st, Pair of Lambs.

SEDBERGH SHOW.—1st, Tup Lamb; 2nd, Tup Lamb; 1st, Three Gimmer Shearlings; 1st, Three Gimmer Lambs; 2nd, Shearling Ram.

HARGREAVES, JAMES,

Laneside, Sawley, Clitheroe.

See Vol. XII., page 40.

Nineteen ewes put to the ram.

Rams used in 1901: Yorkshire Bank 597 and Northern Light 739.

———

HARKER, ANTHONY,

Carperby, Aysgarth Station, R.S.O.

See Vol. XII., page 40.

Eight ewes put to the ram.

Rams used in 1901: Grange Castle Bank 384 and Marquis 786.

———

HINDSON, J. CROSBY,

Hampson Green, Ellel, Lancaster.

See Vol. XII., page 40.

Twenty ewes put to the ram.

Rams used in 1901: Sir George 585 and Rare Luck 654.

———

HITCHON, GILES,

Low Moor, Clitheroe.

See Vol. XII., page 41.

Ten ewes put to the ram.

Rams used in 1901: Norman 648, Asterisk 690 and Dandy Joe 716.

HOGGARTH, JOHN,

Manor Farm, Slyne.

See Vol. XII., page 41.

Twenty-eight ewes put to the ram.

Rams used in 1901: Blood Royal 371, Wolsley 678 and Rob Roy 749.

HOLGATE, JOHN H.,

Foxley Bank, Grindleton, near Clitheroe.

See Vol. XII., page 42.

Ten ewes put to the ram.

Two ewes were added from Mr. Robert Ewbank, one by Hesperus 99, the other by a ram bred by Mr. Rhodes.

Rams used in 1901: Wagtail 766 and Winter 769.

HUTCHINSON, RICHARD,

Brough Sowerby Lodge, Kirkby Stephen.

See Vol. XII., page 42.

Six ewes put to the ram.

Ram used in 1901: Prince Robertus 745.

JACKSON, EDWARD SEDDALL, M.D.,

Carnforth.

See Vol. XII., page 42.

Nine ewes put to the ram.

Ram used in 1901: Slyne Hall 661.

JACKSON, THOMAS,

Netherbeck, Carnforth.

See Vol. XII., page 42.

Forty-five ewes put to the ram.

Rams used in 1901: Chamberlain 537, Marengo 499 and Cathedral 711.

Prizes won in 1901:—

LANCASTER SHOW.—1st and c., Ram lambs; 1st, Gimmer Lambs.

BURTON, MILNTHORPE AND CARNFORTH SHOW.—Special 1st, Three Ram Lambs; 1st, Gimmer Lambs.

JORDAN BROS.,

The Granary, Kendal.

See Vol. XII., page 43.

Rams used in 1901: Baden Powell 602 and Blood Royal II. 606.

JOHNSON, RALPH,

Cragges Farm, Dent, Sedbergh.

See Vol. XII., page 43.

Rams used in 1901: Signet 660 and Helmside 727.

KELLETT, THOMAS,

Slyne-with-Hest, Lancaster.

See Vol. XII., page 43.

Four ewes put to the ram.

Rams used in 1901: A Propos 526, Royal Duke 752.

KETTLEWELL, ROBERT F.,
Camshouse, Bainbridge, Askrigg, R.S.O.
See Vol. XII., page 43.

Six ewes put to the ram.

Rams used in 1901 : Major Curly 443 and Sir Frederick 464.

———

KEY, T. W.,
Old Hall Farm, Casterton, Kirkby Lonsdale.

Eighteen ewes put to the ram.

This flock was begun by purchasing sixteen ewes from Mr. R. Capstick, Bramhaw, Sedbergh: five got by Marthwaite Swell 114, dams by Beatam 9, gr. d. by Lord John 36; five got by Leyburn I. 441, dams by Scargill 124, gr. d. by Beatam 9; six shearlings by Brutus 474, dams by Masterpiece 304, gr. d. by Scargill 124; also two ewes bred by self from ewes descended from Mr. W. Gibson's stock, Killington, and got by a ram bred by Mr. T. Sedgwick, Holme.

Ram used in 1901 : Casterton 708.

———

KITCHEN, THOMAS,
Tatham Hall, Wray, Lancaster.
See Vol. XII., page 44.

Ten ewes put to the ram.

Ram used in 1901: Bellmontina 604.

Prizes won in 1901 :—

BENTHAM SHOW.—1st, Shearling Gimmers; 2nd, Ewes; 1st, Tup Lamb; 2nd, Gimmer Lambs; Special, Collection of Sheep.

KNOWLES, C.,

Sellerly Farm, Ellel, Lancaster.

Twenty-seven ewes put to the ram.

This flock was started many years ago by purchasing three shearling ewes from Mr. W. Blacow. Tups have been used since, bred by Mr. Sedgwick, Sedbergh; Mr. Cragg, Arkholme; Mr. Dargue, Beaumont Grange; and Mr. Thompson, Hest Bank. In 1895 five ewes and eight gimmer lambs were purchased from the late Mr. R. Burrow, Wrayton Hall.

Ram used in 1895: Tatham Swell 275.
,,   1896:   do.
,,   1897: Waterloo 412.
,,   1898:   do.
,,   1899:   do.
,,   1900:   do.
,,   1901: Boy in Blue 472.

MILLNER, WILLIAM and ROBERT,

Slyne Hall, Lancaster.

See Vol. XII., page 44.

Thirty ewes put to the ram.

Rams used in 1901: Harrogate Lad 433, A Propos 526, Rob Roy 749 and Royal Duke 752.

Prizes won in 1901:—

LANCASTER SHOW.—2nd and Reserve, Ram Lambs; 1st, Three Ram Lambs; Reserve for President's Prize for Collection of Sheep.

BURTON, MILNTHORPE AND CARNFORTH SHOW.—1st, Ram Lambs; 2nd, Pair of Ram Lambs; 2nd (Lord Henry Bentinck's Prize), Three Ram Lambs.

KENDAL SHOW.—1st, Ram Lamb; 2nd, Pair of Ram Lambs.

MILLNER, RICHARD,

Bolton-le-Sands, Carnforth.

See Vol. XII., page 44.

Ram used in 1901 : Leading Star 731.

MOFFAT, JOHN,

Rash Farm, Dent, Sedbergh.

See Vol. XII., page 44.

Six ewes put to the ram.

Rams used in 1901 : Erl King II. 485, Blue Boy 696 and British Ruler 705.

MOORE, JOHN,

Yorescott, Askrigg, R.S.O.

See Vol. XII., page 45.

Ten ewes put to the ram.

Ram used in 1901 : Blue Jack 697.

Prizes won in 1901 :—

LEYBURN.—2nd, Three Gimmer Lambs.

HAWES SHOW.—2nd, Three Ewes; 2nd, Three Shearlings; 2nd, Three Gimmer Lambs.

NEWHOUSE, EDWARD,

Ancliffe Hall, Slyne, near Lancaster.

See Vol. XII., page 45.

Thirty-two ewes put to the ram.

Two ewes added from Messrs. W. and R. Millner, one by Regulator 222, and one by Advancer 323.

Rams used in 1901: Royal Ruler 459, Harrogate Lad 438 and Alert 685.

———

NEWTON, NICHOLAS, Senr.,

Overtown Farm, Kirkby Lonsdale.

See Vol. XII., page 45.

Twenty ewes put to the ram.

Rams used in 1901: Royal Blood 656 and Lord Kitchener 734.

———

PARK, J. and T.,

Belmount, Hest Bank, Lancaster.

See Vol. XII., page 46.

Sixteen ewes put to the ram.

Ram used in 1901: Carnforth Swell 706.

———

PARKIN, JOHN,

Sleagill, Penrith.

See Vol. XII., page 46.

Nine ewes put to the ram.

Rams used in 1901: Peggleside 395 and Umpire 409.

PARKINSON, RICHARD,

Bennett's Farm, Preesall.

Ten ewes put to the ram.

This flock was commenced by purchasing ten ewes from Mr. T. Jackson, Netherbeck, Carnforth: two got by Cashier 327, two by Lord of the Valley 109, four by Cardinal 477, and two by Boy in Blue 472.

Ram used in 1901: Ashbourne 687.

———

PERCIVAL, JOHN,

Carperby, Aysgarth Station, R.S.O.

See Vol. XII., page 42.

Sixteen ewes put to the ram.

Rams used in 1901: Shepherd's Delight 659 and Medalist 737.

Prizes won in 1901:—

WENSLEYDALE AGRICULTURAL SHOW AT LEYBURN.—2nd, Ram Lamb; 2nd, Best Three Wensleydale Sheep.

HAWES.—1st and 2nd, Ram Lambs; Society's Silver Medal for Best Pair of Lambs.

———

PICKERING, L. T.,

The Hall Farm, Bourn, Cambridge.

See Vol. XII., page 47.

Forty-two ewes put to the ram.

Ram used in 1901: Le Roi 640.

PROCTER, JOHN,

Eccles Farm, Westhouse, Kirkby Lonsdale.

See Vol. XII., page 47.

Two ewes put to the ram.

Ram used in 1901: Credit 715.

———

PROCTER, R.,

Oak Mount, Burnley, and Barkerfield, Downham, Clitheroe.

See Vol. XII., page 47.

Seventeen ewes put to the ram.

Rams used in 1901: Maximillian 645 and Peter Teazle 741.

———

REDMAN, RICHARD,

The Brows, Glasson Dock, Lancaster.

Ten ewes put to the ram.

This flock was started by ewes purchased in the first instance from Mr. Hutchinson, Gowan Hall, Kirkby Lonsdale. One ewe has since been added bred by Mr. A. Ewan, Gooda. Rams have been used bred by the following breeders: Mr. Rhodes, Westhouse (2), Mr. J. Hoggarth, Slyne, and the late R. Burrow, Wrayton Hall (2).

Ram used in 1900: Waterloo 412.

,,   1901: Boy in Blue 472 and Apollo 686.

RENSHAW, J. E.,

Toulbrick Farm, Hambleton, Poulton-le-Fylde, and
Lark Hill, Walmersley Road, Bury.

See Vol. XII., page 47.

Twelve ewes put to the ram.

Four shearling ewes were added in 1901—two from Mr.
Burra, one of which was by Erl King II. 485, dam by
Wellington 236, gr. d. by Regulator 222; the other by
Erl King II. 485, dam by Recollection 453, gr. d. by
Wellington 236, g. gr. d. by Lord of the Valley 109; and
two from Mr. Handley, sire Powder Blue 397, dam by
Scarbank 271.

Prizes won in 1901:—

GARSTANG.—2nd, Aged Ram ; 2nd, Ram Lamb.

GOOSNARGH—1st, Ram Lamb.

RHODES, WILLIAM,

Lundholme, Westhouse, Kirkby Lonsdale.

See Vol. XII., page 48.

Twenty-six ewes put to the ram.

Rams used in 1901: Marengo 499, Blue Beard 607
and Royal Maidstone 582.

RIGG, REDMAYNE,

Wells House, Cartmel, Lancashire.

See Vol. XII., page 49.

Fifteen ewes put to the ram.

Added five ewes from Mr. W. Gibson by Stainton 405, also three gimmer lambs by Successor 514.

Rams used in 1901 : Dainty Stamp 621, Cast Out 709.

Prizes won in 1901 :—

NORTH LONSDALE.—2nd, Shearling Ram, DAINTY STAMP 621 (open class) ; 2nd, local class ; 1st, Breeding Ewes ; 1st, Shearling Ewes.

CARTMEL.—1st, Shearling Ram, DAINTY STAMP 621 ; 2nd, Ram Lamb ; 1st, Shearling Ewes ; 2nd, Gimmer Lambs ; 1st, best Three Sheep, any breed ; 1st, silver cup, Best Collection ; 1st, best Female Sheep.

ROBINSON, H. and G. G.,

Strickley Farm, Old Hutton, Kendal.

See Vol. XII., page 49.

Ten ewes put to the ram.

Rams used in 1901: Wonder 679 and Blue Beard 601.

ROBSON, A.,

Carperby, Aysgarth Station, R.S.O.

See Vol. XII., page 49.

Six ewes put to the ram.

Ram used in 1901 : Jester 728.

SCARR, JAMES,

Colby Hall, Bainbridge, Askrigg, R.S.O.

See Vol. XII., page 50.

Six ewes put to the ram.

Ram used in 1901 : Lord Roberts 568.

SEDGWICK, THOMAS,

Low Holme, Middleton, Sedbergh.

See Vol. XII., page 50.

Nineteen ewes put to the ram.

Ram used in 1901 : Better Luck 698.

Prizes won in 1901 :—

SEDBERGH.—1st, Shearling Ram.

KIRKBY LONSDALE.—2nd, Ram Lamb.

SOCIETY'S SHOW AND SALE AT HELLIFIELD.—1st for Shearling Ram, and Silver Medal for Best Animal.

SEDGWICK, JOHN,

Low Holme, Sedbergh.

See Vol. XII., page 50.

Fourteen ewes put to the ram.

Ram used in 1901: Better Luck 698.

SIMPSON, RICHARD,

Manor Farm, Bletchley, Bucks.

See Vol. XII., page 51.

Nineteen ewes put to the ram.

Ram used in 1901 : Clare Pride 718.

STAINTON, JOHN,

Hallbeck, Killington, Kirkby Lonsdale.

See Vol. XII., page 51.

Six ewes put to the ram.

One ewe added from Mr. T. Woof, Borrett, by William Tell 281, dam by Borrett 147, gr. d. by Freeman 155.

Ram used in 1901 : Estimation 487.

STOREY, H. J.,

Bailrigg, Lancaster.

Six ewes put to the ram.

In 1900 six ewes were purchased from Mr. Jos. Towers, Nether Kellet; four were got by Viking 278 and two by Lancaster Fashion 210, dams by Dalesman 22, gr. d.'s by Bread Baker 372 and Limestone Lad 389.

Ram used in 1901 : Bridegroom 534.

STUART, RICHARD,

Brock Vale, Sowerby, Garstang.

See Vol. XII., page 51.

Seventeen ewes put to the ram.

Two gimmer lambs added purchased from Mr. Renshaw.

Ram used in 1901: Blue Skin 532.

Prizes won in 1901 :—

LYTHAM.—1st, Ewes; 1st, Ram; 1st, Shearling Gimmers; 1st, Gimmer Lambs; 3rd, Ram Lamb.

BLACKPOOL (OPEN).—3rd, Ram Lamb; 2nd and 3rd, Ewes; 2nd, Shearling Gimmers; 2nd, Gimmer Lambs.

GARSTANG.—1st and 2nd, Ewes; 1st, Shearling Gimmers; 1st, Gimmer Lambs.

GARSTANG AUCTION MART.—1st, Ram Lamb.

GOOSNARGH (OPEN).—1st, Ram; 1st and 3rd, Ram Lamb; 1st, Ewes; 1st and Special, Shearling Gimmers; 1st, Gimmer Lambs.

PRESTON AUCTION MART.—2nd, Shearling Ram; 3rd, Ram Lamb.

BROCKHOLES AUCTION MART.—2nd, Shearling Ram.

———

TAYLOR, C. EDWARD,
Akay, Sedbergh.
See Vol. XII., page 52.

Three ewes put to the ram.

Rams used in 1901: Powder Blue 397 and Beauty 603.

———

THOMPSON, THOMAS,
Prospect House, Hest Bank, near Lancaster.
See Vol. XII., page 53.

Twenty-one ewes put to the ram.

Three ewes added from Mr. R. Winder, sire Falconer 297.

Rams used in 1901: Prospect King 746, Royal Count
751 and Prospect Viking 747.

Prizes won in 1901 :—

BLACKPOOL.—1st, Aged Ram, PEARL KING; 1st and 2nd, Shearling Ram; 1st Ram Lamb; 1st, Three Ewes; 1st, Three Shearling Ewes; 1st, Three Ewe Lambs.

LANCASTER.—2nd, Aged Ram, PEARL KING; 2nd, Shearling Ram; 1st, Three Ewes; 3rd, Three Shearling Ewes; 3rd, Three Ewe Lambs.

MILNTHORPE, BURTON AND CARNFORTH.—2nd and 3rd, Aged Ram, PEARL KING and PROSPECT VIKING; 2nd, Ram Lamb; 2nd, Three Ewes.

KENDAL.—2nd, Aged Ram, PEARL KING; 1st and 2nd, Shearling Ram; 2nd and 3rd, Ram Lamb; 1st, Three Ewes; 3rd, Three Shearling Ewes; 1st, Three Ewe Lambs; Silver Medal, Pair of Wensleydale Lambs; 2nd, Collection.

———

TOWERS, FRANCIS,

Stubb Hall, Nether Kellet, Carnforth.

Twenty ewes put to the ram.

The flock was founded in 1887 by purchasing five ewes from Messrs. W. and R. Millner, Slyne Hall, Lancaster, which were served by one of their rams.

Ram used in	1888 :	Limestone Lad 389.	
,,	1889 :	do.	
,,	1890 :	do.	Bread Baker 372.
,,	1891 :	do.	do.
,,	1892 :	Bread Baker 372, Dalesman 22.	
,,	1893 :	do.	do.
,,	1894 :	Danegelt 717.	
,,	1895 :	do.	Tinpocket 517.
,,	1896 :	Tinpocket 517.	
,,	1897 :	do.	
,,	1898 :	do.	Viking 278.
,,	1899 :	Estimate 486.	
,,	1900 :	do.	
,,	1901 :	do.	Harrogate Lad 433.

TOWERS, JOSEPH,

Lawson's Farm, Nether Kellet, Carnforth.

See Vol. XII., page 53.

Forty-three ewes put to the ram.

Ten ewes added from Mr. T. Thompson, Hest Bank.

Rams used in 1901: Bridegroom 534, Best Man 692.

Prizes won in 1901 :—

SOCIETY'S SHOW AND SALE AT HELLIFIELD.—2nd, 6th, Reserve and h.c. for Ram Lamb.

———

WALKER, THOS.,

Templand, Grange-over-Sands.

Thirteen ewes put to the ram.

This flock has been in existence a considerable time, the ewes being obtained from Mr. Swindlehurst, Docker. Rams have been used bred by members of the Society, viz., Messrs. T. Jackson, J. Dargue, J. Gibson, S. Bargh, and the late T. Willis.

Ram used for 3 years: Templand Boy 758.
,,   ,, Warton 77.
,,   ,, Templand Gem 760.
,, in 1900: Templand Chief 759.
,, in 1901: Templand Swell 761.

———

WALLER, JOHN,

Low Thoresby, Redmire, Aysgarth Station, R.S.O.

See Vol. XII., page 54.

Forty-five ewes put to the ram.

Ram used in 1901: Long Tom 733.

WALSH, J., The Exors. of the late,

Bourne Hall, Poulton-le-Fylde.

See Vol. XII., page 54.

Seventeen ewes put to the ram.

Six ewes added from Mr. W. Rhodes: three by Wellington 236 and one each by True Blue 238, Royal Blue 350 and Erl King 382.

Rams used in 1901: Bourne Champion 611 and Top Sawyer 762.

Prizes won in 1901 :—

BLACKPOOL.—2nd, Aged Ram ; 3rd, Gimmer Lambs.

PRESTON.—1st, Aged Ram ; 2nd, Ram Lamb.

POULTON-LE-FYLDE.—1st, Aged Ram ; 3rd, Ram Lamb.

WHITTAKER, G.,

Gibson's Farm, Kirkland, Garstang.

Twenty-four ewes put to the ram.

This flock was founded some sixteen years ago by purchasing four ewes from Mr. A. Ewan, Gooda, which were served by Carperby's Best 14. Eight years ago three gimmer lambs were purchased from Mr. Ewan. Pure-bred rams were always used, bred by Mr. J. Shepherd, Churchtown, Mr. Ewan, Dr. Jackson and others.

Ram used in 1894: Dark Blue 483.

,,   1895:   do.

,,   1896: Masterful 303.

,,   1897:   do.

,,   1898: Blue Bell 418.

,,   1899:   do.

,,   1900: Yorkshire Fashion 413.

,,   1901:   do.

WHITWELL, JOHN,

Holden House, Westhouse, Kirkby Lonsdale.

See Vol. XII., page 54.

Eight ewes put to the ram.

Rams used in 1901: Estimation 487, New Boy 740.

WILLIS, THOMAS, The Exors. of the late,

Manor House, Carperby, Aysgarth Station, R.S.O.

See Vol. XII., page 55.

Forty ewes put to the ram.

Rams used in 1901: Estimation 487, Royal Maidstone 582, Royal York 658, Royal Cardiff 750 and Royal Scotsman 753.

Prizes won in 1901:—

ROYAL SHOW AT CARDIFF.—1st, £10, and 2nd, £5, Shearling Ram; 1st, £10, and 2nd, £5, Three Shearling Ewes.

GREAT YORKSHIRE SHOW AT BRADFORD.—1st, £7, Aged Ram, ROYAL YORK 658; 1st, £10, and 2nd, £5, Shearling Ram; 1st, £3, and Reserve No., Ram Lamb; 1st, £7, and 2nd, £5, Three Shearling Ewes.

ROYAL LANCASHIRE SHOW AT ST. HELENS.—1st, Aged Ram, ROYAL YORK 658; 1st, and 2nd, Shearling Ram; 2nd, Ram Lamb; 1st and 2nd, Three Shearling Ewes; Champion Cup, and Reserve Champion for the Best Ram and Pen of Ewes; Champion Cups for the Best Ram Lamb, and Pen of Ewe Lambs.

WENSLEYDALE AGRICULTURAL SOCIETY'S SHOW (DISTRICT CLASSES).—1st, Aged Ram, ESTIMATION 487; 1st and Reserve No., Shearling Ram; 1st and 3rd, Ram Lamb; 1st, Three Shearling Rams suitable for crossing purposes; 1st, Three Ewes.

WENSLEYDALE AGRICULTURAL SOCIETY'S SHOW (OPEN CLASSES).—1st and 2nd, Shearling Ram; 2nd, Ram Lamb; 1st and 2nd, Three Shearling Ewes.

CRAVEN SHOW AT SKIPTON.—1st, Aged Ram, ROYAL YORK 658; 1st and 2nd, Shearling Ram; 1st, Three Ewes; 1st and 2nd, Three Shearling Ewes; 1st, Three Ewe Lambs.

WILLS, JAMES,

Garth Heads Road, Appleby.

See Vol. XII., page 56.

Twelve ewes put to the ram.

Rams used in 1901: President 578, The Swell 765, and Dufton Royal 718.

WINDER, ROBERT,

Townside Farm, Pilling, Garstang.

See Vol. XII., page 56.

Ten ewes put to the ram.

Ram used in 1901 : Dutiful Boy 719.

WOOF, THOMAS,

Borrett, Sedbergh.

See Vol. XII., page 57.

Twenty-four ewes put to the ram.

Rams used in 1901 : Estimation 487, Gibson 725.

Prizes won in 1901 :—

KIRKBY LONSDALE.—Special Prize for Aged Ram.

SEDBERGH.—1st, Aged Ram ; 1st and Reserve, Ram Lambs (selling class).

SOCIETY'S SHOW AND SALE AT HELLIFIELD.—1st, Aged Ram ; Reserve, Shearling Ram ; Bronze Medal for Second best Animal.

CHARACTERISTIC POINTS

OF

WENSLEYDALE BLUE-FACED SHEEP.

Wool—Bright and lustrous. Flat staple of medium breadth and good length; each staple curled or pirled out to the end. Of equal staple all over the back and sides from shoulder to breech. The whole free and open, and free from mistiness on the back. When the sheep is turned, the belly, and particularly the scrotum, should be well covered with wool and free from hair.

Head—Broad at the muzzle, especially in rams. Back of head flat and wide between ears. The face seen in profile should show good depth of jaw. Ears of good size, neatly set on and well carried. Head and ears of a deep blue tinge, which often extends to the rest of the body. Entire absence of hair about the forehead, back of head and ears. Tuft of wool on forehead. Back of head, especially round the ear roots, covered with fine wool. Free from *coarse* hair on the rest of the face. Eyes bright and full.

Neck—Of good length, and strong, rising gracefully from the shoulders, and carrying the head a good height.

Shoulders and Crops—Shoulders well laid back into the crops, which should be wide and full.

Chest—Coming well down and forward between the fore-legs, and wide on the floor of the chest and hockster.

Back, Loins, Sides and Quarters—Ribs well sprung, deep, and great length of side. Loins broad and well covered with firm flesh along the back. Hind-quarters long, square and nicely packed. Tail broad.

Thighs, Legs and Feet—Thighs well down to the hock, large and broad behind. Twist full. Legs with plenty of bone, but freedom from coarse hair, straight set on at each corner and well apart. Hind legs with a nice covering of fine wool from hock to hoof. Feet moderately large and well formed.

SCALE OF POINTS.

Wool	20
Head	20
Neck	10
Shoulders and Crop	10
Chest	10
Back, Loins, Sides and Quarters	20
Thighs, Legs and Feet	10
Total	100

LIST OF MEMBERS.

Ashton, Lord, Ryelands, Lancaster.

Abbot Bros., Thuxton, Norfolk.

Bainbridge, Thos., Brough Castle, Kirkby Stephen.

Banks, J. H., Trunnah Farm, Thornton, Poulton-le-Fylde.

Banks, Thomas, Bilsborough, Preston, Lancashire.

Bargh, Samuel, Clifford Hall, Burton-in-Lonsdale, Kirkby Lonsdale.

Barker, J., Maythorne, Leyburn.

Bateman, John, Hallgarth, Clapham, Yorkshire.

* Bentinck, Lord Henry, M.P., Underley Hall, Kirkby Lonsdale.

Bleasdale, Joseph, Pump House, Nether Kellet, Carnforth.

* Burra, R., Gate, Sedbergh, R.S.O.

* Burra, R., Junr., Gate, Sedbergh, R.S.O.

Camplin, J. T., Newby, Penrith.

Capstick, Richard, Bramhaw, Sedbergh.

Capstick, R. and Son, Moser Hill, Dent, Sedbergh.

Chayter, Sir W. H. E., Croft Hall, Darlington.

Close, James, Carperby, Aysgarth Station, R.S.O.

Cock, E. and Sons, Red Bank, Bolton-le-Sands, Carnforth.

Cooper, W. and Nephews, Berkhamsted.

Croft, J. Carter, Carperby, Aysgarth Station, R.S.O.

Dargue, J., Bee Nest, Casterton, Kirkby Lonsdale.

Dent, W. and Sons, The Street, Bolton, Penrith.

Dickenson, John, Pit Farm, Grange-over-Sands.

Dinsdale, R., Redmire, Leyburn, R.S.O.

Dinsdale, W., Junr., Carperby, Aysgarth Station, R.S.O.

Domoney, J. W., Sedbergh.

Dormer, Mrs. Upton Cottrell-, Ingmire Hall, Sedbergh.

Ewan, Aaron, Gooda, Westhouse, Kirkby Lonsdale.

Ewbank, Robert, Lawkland Green, Clapham, Yorkshire.

Farrer, J. A., Ingleboro', Clapham, Yorkshire.

Farrer, Mrs. J. A., Clapham, Yorkshire.

Farrer, John, Abbot's Reading, Haverthwaite, Ulverston.

Fawcett, T., Holme, Carnforth.

* Fitzwilliam, Hon. W. H. W., Wigganthorpe, York.

* Fothergill, J. W., Brownber, Newbiggin, R.S.O., Westmorland.

Foster, Col. W. H., Hornby Castle, Lancaster.

Garth, F., Haverdell House, Low Row, Richmond.

Gibson, R. W., Orton, Tebay, Westmorland.

Gibson, J. and Son, 163, Highgate, Kendal.

Gibson, W., Broadrain, Killington, Sedbergh.

Greene, Captain H. Dawson, Whittington Hall, Kirkby Lonsdale.

Graham, W., Eden Grove, Bolton, Penrith.

Handley, John, Brigflatts, Sedbergh.

Hargreaves, James, Lane Side, Sawley, Clitheroe.

Harker, Anthony, Carperby, Aysgarth Station, R.S.O.

Helme, Norval W., M.P., Springfield, Lancaster.

Hindson, J. C., Hampson Green, Ellel, Lancaster.

Hitchon, Giles, Low Moor, Clitheroe.

Hoggarth, J., Manor Farm, Slyne.

Holgate, John H., Foxley Bank, Grindleton, near Clitheroe.

Hornby, Major E. G. S., Dalton Hall, Burton, Westmorland.

Hutchinson, Richard, Brough Sowerby Lodge, Kirkby Stephen.

Jackson, E. S., M.D., Carnforth.

Jackson, Thomas, Netherbeck, Carnforth.

Johnson, Ralph, Cragges, Dent. Sedbergh.

Jordan Bros., The Granary, Kendal.

Kellett, T., Slyne-with-Hest, Lancaster.

Kettlewell, R., Camshouse, Bainbridge, Askrigg.

Key, T. W., Old Hall Farm, Casterton, Kirkby Lonsdale.

Kitchen, Thomas, Tatham Hall, Wray, Lancaster.

Knowles, C., Sellerly Farm, Ellel, Lancaster.

* Lowther, Hon. W., Lowther Lodge, Kensington Gore, London, W.

Millner, Richard, Bolton-le-Sands, Carnforth.

Millner, W. and R., Slyne Hall, Lancaster.

Moffat, John, Rash Farm, Dent, Sedbergh.

Moore, John, Yorescott Farm, Askrigg, R.S.O.

Newhouse, E., Ancliffe Hall, Slyne, Lancaster.

Newton, N., Senr., Overtown Farm, Kirkby Lonsdale.

Park, J. and T., Belmount, Hest Bank, Lancaster.

Parkin, John, Sleagill, Penrith.

Parkinson, R., Bennett's Farm, Preesall.

Percival, John, Carperby, Aysgarth Station, R.S.O.

Pickering, L. T., Bourn, Cambridge.

Procter, J., Eccles Farm, Westhouse, Kirkby Lonsdale.

Procter, R., Oak Mount, Burnley, and Barkerfield, Downham, Clitheroe.

Punchard, F., Kirkby Lonsdale.

Redman, R., The Brows, Glasson Dock, near Lancaster.

Renshaw, J. E., Toulbrick Farm, Hambleton, Poulton-le-Fylde, and Larkhill, Walmersley Road, Bury.

Rhodes, William, Lundholme, Westhouse, Kirkby Lonsdale.

Rigg, R., M.P., Applegarth, Windermere.

Rigg, Redmayne, Wells House, Cartmel, Lancashire.

Robinson, H. and G. G., Strickley Farm, Old Hutton, Kendal.

Robson, A., Carperby, Aysgarth Station, R.S.O.

Scarr, James, Colby Hall, Bainbridge, Askrigg, R.S.O.

Sedgwick, Thomas, Low Holme, Middleton, Sedbergh.

Sedgwick, John, Low Holme, Middleton, Sedbergh.

Simpson, Richard, Manor Farm, Bletchley, Bucks.

Stainton, J., Hallbeck, Killington, Kirkby Lonsdale.

Storey, H. J., Bailrigg, Scotforth, Lancaster.

Stuart, Richard, Brock Vale, Sowerby, Garstang.

Tatham, G. S., Kirkby Lonsdale.

Taylor, C. E., Akay, Sedbergh.

Thompson, F. Whitley, M.P., Saville Heath, Halifax.

Thompson, T., Prospect, Hest Bank, Lancaster.

Thompson, S., Prospect, Hest Bank, Lancaster.

Towers, Joseph, Lawson's Farm, Nether Kellet, Carnforth.

Towers, Francis, Stubb Hall, Nether Kellet, Carnforth.

Waller, John, Low Thoresby, Aysgarth Station, R.S.O.

Walker, Thos., Templand, Grange-over-Sands.

Walsh, John, Exors. of the late, Bourne Hall, Poulton-le-Fylde.

Whittaker, George, Gibson's Farm, Kirkland, Garstang.

Whitwell, John, Gill Foot, Mansergh, Kirkby Lonsdale.

Wilcock, William, Ravensclose, Wennington, Lancaster.

Willis, Thos., Exors. of the late, Manor House, Carperby, Aysgarth Station, R.S.O.

Willis, John, East End House, Carperby, Aysgarth Station, R.S.O.

Wills, James, Nether Hoff, Appleby, Westmorland.

Winder, Robert, Townside Farm, Pilling, Garstang.

Woof, T., Borrett, Sedbergh.

* Those gentlemen whose names are marked with an asterisk are Life Members of the Society.

END OF VOL. XIII.

PRINTED BY T. HISCOCK, THE WENSLEYDALE PRESS, HAWES.

Ness & Company's Sheep Dip

Is the true friend of the Flockmaster, helping the Sheep to thrive. By its action upon the Skin and the Wool it aids in GAINING PRIZES, and it is used and liked by some of the most important Prize Winners of the North.

TESTIMONIALS.

Mr. RAWLINSON, Docker Hall, Kendal.

"19th Jan., 1897. . . . We like it the best of all the dips ever used by us."

"18th May, 1897. . . . We still like your Dip, and are sending two Rams to the Royal Show at Manchester, dipped with your wash previous to going."

Mr. T. RAWLINSON, Park House, Kirkby Lonsdale, 24th July, 1899.

"I have used your Dip for 1600 since February, and it I'll have and no other. We have always liked your preparations."

Messrs. JAMES GIBSON & SON, Natland Hall, near Kendal.

"October 10th, 1896.—Have great pleasure in recommending your Dip."

"January 18th, 1897.—Your Dip has done well for us, and the sheep are clear of ticks or any kind of vermin. None we like so well as yours."

Messrs. R. PARKER & SON, Moss End Farm, Milnthorpe, 10th February, 1897.

"We used your Dip last Autumn on 200 sheep, and like it very well. We have also used it amongst CATTLE for LICE, and find it answers well."

Mr. J. H. DENT, Elm House, Bolton, Penrith, 9th September, 1901.

"I am sending you a photo of my Oxford Down Ram, which during the present show season has never been beaten. Having used your Dip I can bear testimony to its excellence."

In Tins at 2s. 6d., 4s., and 7s. 6d.

AGENTS: O. R. BOWE, Hawes; AARON KNAGGS, Grocer, &c., Bainbridge; J. SWINBANK, Chemist, Bedale.

WRITE FOR A SAMPLE TO THE SOLE MANUFACTURERS AND PROPRIETORS,

Ness & Company, Darlington.

Offices: Church Row.

Another Great Success.

AT THE NEW SOUTH WALES

SHEEP BREEDERS' GREAT SHOW

AT SYDNEY, JUNE, 1901,

SHEEP DIPPED WITH

Little's Dips

Secured the only 2 Grand Champion Prizes.

14 Champion Prizes.

24 First Prizes.

21 Second Prizes.

18 Third Prizes.

Total 79 PRIZES.

What other proof do you want that

Little's Dips

are the best for your Sheep?

LITTLE'S DIPS (Fluid, Powder, Paste) give Best Results and are unequalled for increasing growth and improving quality of Wool.

Sold everywhere, or direct from the Manufacturers:

Morris, Little & Son, Ltd.,

Doncaster.

The Executors of the late Thos. Willis,

MANOR HOUSE, CARPERBY,
AYSGARTH, R.S.O.,
BREEDERS AND EXPORTERS OF

Pedigree Wensleydale Blue-faced Sheep.

THIS FLOCK, the oldest of the breed, was not extensively represented in the Show Ring until seven years ago, during which time Shearling Rams have been placed First at the Royal Shows on no fewer than six occasions. At the Royal Show at York in 1900, four firsts and the champion prize for the best ram were won, and last year all the prizes offered for the breed at Cardiff.

In the Sale Ring, the above breeders have been no less successful, their Shearling Rams averaging over £2 per head more than any other lot at the principal ram sale in Scotland.

The use of Carperby rams in pure flocks has always been attended with the highest success, and so many years of careful selection and breeding have made them particularly famous amongst breeders of cross lambs, on account of the fine quality and evenness of their get, and the ability of the rams to stand the most severe hill-work.

The sires used during the past season include:—

ROYAL MAIDSTONE 582 (see opposite page).

ROYAL YORK 685, First Prize Shearling Ram at the Royal Show at York in 1900, and unbeaten as a Two-shear last season.

ROYAL CARDIFF, First Prize Shearling at the Royal Show at Cardiff, and unbeaten throughout the season.

TWO-SHEAR WENSLEYDALE RAM.
Never been beaten.   Weight 448lb.

ROYAL MAIDSTONE 582.

WINNER OF
1899.   1st prize, Royal Show, Royal Lancashire and Great Yorkshire, etc.
1900.   1st and Champion, Royal Show; 1st and Champion, Royal Lancashire;
         1st, Great Yorkshire, etc.

www.ingramcontent.com/pod-product-compliance
Lightning Source LLC
Chambersburg PA
CBHW081740220526
45468CB00008B/2180